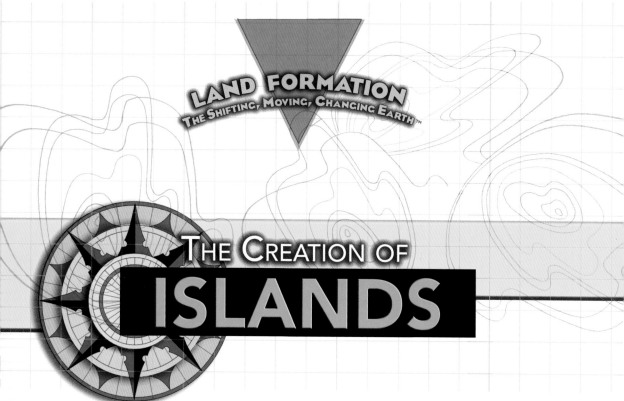

LAND FORMATION
THE SHIFTING, MOVING, CHANGING EARTH™

THE CREATION OF
ISLANDS

Greg Roza

New York

For Kim, Josh, Avery, and Emma

Published in 2010 by The Rosen Publishing Group, Inc.
29 East 21st Street, New York, NY 10010

First Edition

Library of Congress Cataloging-in-Publication Data

Roza, Greg.
The creation of islands / Greg Roza.—1st ed.
    p. cm.—(Land formation: the shifting, moving, changing earth)
Includes bibliographical references.
ISBN-13: 978-1-4358-5299-0 (library binding)
ISBN-13: 978-1-4358-5596-0 (pbk)
ISBN-13: 978-1-4358-5597-7 (6-pack)
1. Plate tectonics. 2. Islands. 3. Geology, Structural. I. Title.
QE511.4.R69 2009
551.42—dc22

2008040847

Manufactured in Malaysia

On the cover: A tropical island.

CONTENTS

INTRODUCTION

An island is an area of land that is completely surrounded by water. Generally speaking, an island is usually a small area of land in comparison to the body of water in which it is found. Some islands are big enough to hold millions of people, while others are barely big enough for a single person to stand on.

The largest island on Earth, Greenland, is 836,330.48 square miles (2,166,086 square kilometers). The Indonesian island of Java is the most highly populated island in the world; approximately 124 million people live there.

Indonesia is a nation of volcanic islands between the Indian Ocean and the Pacific Ocean. It is made up of about 17,508 islands—6,000 of which are inhabited—and has 33,999 miles (54,716 km) of coastline. Indonesia has more active volcanoes than any other country on Earth.

About 1,500 years ago, a large volcanic island between Sumatra and Java erupted so violently that it collapsed to form three separate islands: Verlaten Island; Lang Island; and Krakatoa Island, which was made up of three active volcanoes.

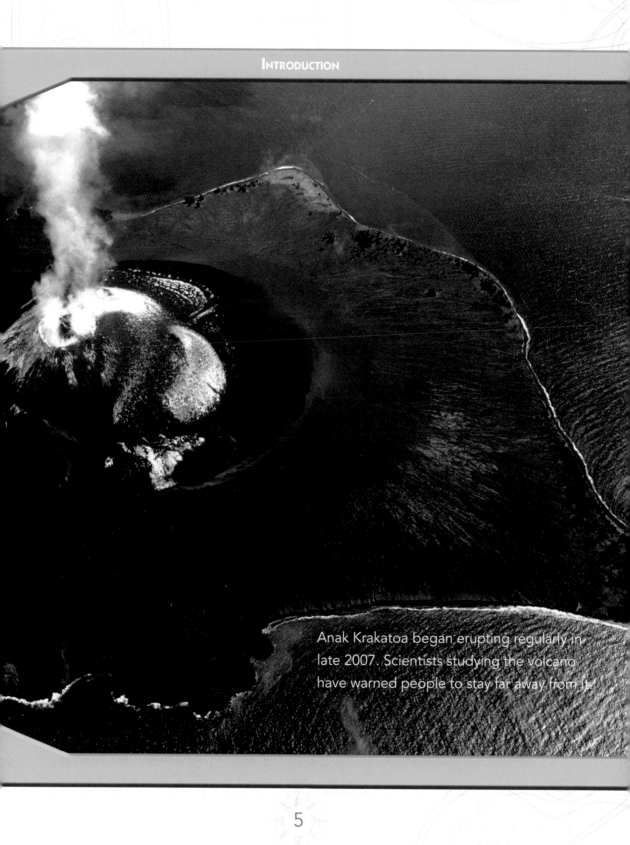

Anak Krakatoa began erupting regularly in late 2007. Scientists studying the volcano have warned people to stay far away from it.

On August 26 and 27, 1883, roughly two-thirds of the island of Krakatoa exploded and sank into the sea. A rain of hot ash killed about a thousand people on Sumatra. The blast also created powerful tsunamis that reached as far away as South Africa. The ash produced by the volcano was so immense that it caused temperatures worldwide to drop by approximately 1.8 degrees Fahrenheit (1 degree Celsius) for several years. Lava continued to pour out of the remains of Krakatoa until a new island rose above the surface of the ocean. This volcano is called Anak Krakatoa, which means "Child of Krakatoa." It continues to grow today. Scientists wonder if Anak Krakatoa will one day explode much like its parent, Krakatoa.

The tumultuous history of Krakatoa emphasizes the powerful geological forces needed to make and destroy islands. However, the construction and destruction of islands rarely occurs as quickly as this example. Many islands require millions of years to reach maturity.

In this book, we will examine the way that Earth creates and destroys islands. We will also learn about many interesting island nations and the ways people have learned to make their own islands.

EARTH: THE ISLAND-MAKER

There are so many islands on Earth that it is difficult to say exactly how many there are. Scientists estimate that there are more than 100,000 islands on our planet, many of which are inhabited by people. Islands can be found in the middle of a vast ocean or in a lake. They can occur as solitary patches of land or in groups numbering in the thousands.

Due to Earth's powerful forces, islands are changing all the time. Volcanic activity and other forces cause islands to grow. Erosion in the form of ocean tides and landslides causes them to shrink. Many islands, especially volcanic islands, eventually disintegrate and crumble into the ocean. Some disappear when the sea level rises, although many form in the same way.

Earth's islands were not all made the same way. To truly understand the different kinds of islands on Earth, as well as how they form and vanish, it is important to understand our planet's "island-forming" forces.

Continental Drift

Earth's land may seem solid and immovable, but it is actually moving all the time— just very slowly. It has been doing this for

EARTH'S 10 LARGEST ISLANDS

Island	Size
Greenland	836,330 square miles (2,166,086 square kilometers)
New Guinea	303,381 sq miles (785,753 sq km)
Borneo	288,869 sq miles (748,168 sq km)
Madagascar	226,917 sq miles (587,713 sq km)
Baffin	194,574 sq miles (503,944 sq km)
Sumatra	171,069 sq miles (443,066 sq km)
Honshu	88,982 sq miles (225,800 sq km)
Great Britain	88,787 sq miles (229,957 sq km)
Victoria	85,154 sq miles (220,548 sq km)
Ellesmere	71,029 sq miles (183,965 sq km)

millions of years. In fact, at one time nearly all of the land on Earth was gathered together in a single landmass that scientists call Pangaea. Over the course of millions of years, pieces of Pangaea gradually drifted away from each other. Very slowly, the landmasses moved into the positions that we are familiar with today.

When we look at a map of the continents today, it is possible to see how some of the original pieces of Pangaea fit together. Perhaps the most obvious remnants of that early landmass are the shapes of the east coast of South America and the west coast of Africa. At one time, many millennia ago, these two landmasses fit together like two pieces of a puzzle. Rocks found in Namibia, Africa, are very similar to those found in Brazil, South America, further verifying the fact that the two continents used to be connected.

Continental drift is not some ancient event that ended many years ago. It is still happening right now, and it will continue to happen as long as Earth keeps making energy in its core, or center. Many millions of years from now, the continents will likely collide once again to form a new supercontinent—Pangaea II.

You might be wondering how it is possible that Earth's giant landmasses could move at all. They seem to be so vast and so permanent. What are the forces at work that allow continents to slide over Earth's surface like oil patches on water? The answer is plate tectonics.

Plate Tectonics

Earth is made up of several layers. The top layer, the one on which we live, is called the crust. The crust is made up of solid rocks and minerals, and much of it is covered with water. In most places, the crust is about 25 miles (40 km) thick. The crust at the bottom of the ocean may be as thin as 5 miles (8 km). While Earth's crust may seem like a solid layer, it is actually made up of plates. There are seven major plates and many other smaller plates. Plate tectonics deals with the way these plates interact with each other.

There are several theories about how plate tectonics works. Most scientists think that Earth's plates "float" on a softer, hotter layer of rock beneath them. Beneath the crust is a much thicker layer called the mantle. Together, the crust and the outermost layer of the mantle are Earth's lithosphere.

The next layer is known as the asthenosphere. It is weaker and softer than the rigid plates of the lithosphere. The bottom of the asthenosphere is hotter than the top. Just as hot air rises and cold

air sinks, the heat of the lower mantle rises very slowly, causing the upper layer of the mantle to move very slowly—only a few centimeters a year. Many scientists believe that this heating action causes the asthenosphere to act somewhat like a body of water beneath the lithosphere. Tectonic plates are able to drift very slowly over this softer layer.

Plates Collide

There are no gaps between tectonic plates. The borders where the plates meet are centers of great force. This is where mountains rise, earthquakes rumble, and volcanoes explode. We also sometimes find islands on or near these borders. Different things

There are different types of plate boundaries. The plates on the far left are moving away from each other, and new land is forming between them. In the center, one plate is being forced beneath another plate.

can happen depending on how two plates interact. One plate may slide under or over another. Two plates may also drift apart, or they may move past each other.

Making Mountains

It may seem very hard to believe, but as two tectonic plates push against each other, they can bunch up like two pieces of paper. As the plates continue to push into each other, the land rises higher and higher. After millions of years, this action can give birth to mountains. To better understand this idea, one can look at Asia.

Long ago in Earth's past, the land that is today known as India was separate from Asia. When Pangaea first broke up, India was

The shapes of modern-day landmasses, particularly Africa and South America, are distinguishable in the ancient supercontinent known as Pangaea.

somewhat like an island drifting north more quickly than other landmasses—about 1 inch (2.5 centimeters) a year. This landmass collided with Asia about 55 million years ago and kept speeding north. India's landmass was forced beneath Asia, which caused the land above the collision to rise dramatically. This process has continued since the collision and will continue for many years to come. The result is the largest mountain range in the world, the Himalayas. India will continue to sink below Asia, and the Himalayas will continue to rise ever higher.

Volcanic Activity

The friction between the plates often causes earthquakes. It also causes rocks to melt and rise to the surface in the form of magma. When the magma reaches Earth's surface, it is called lava. The lava spills out over the land and hardens to form new rocks. As this happens again and again over the course of many years, the new rocks form dome- and cone-shaped landforms called volcanoes. This may happen on land or in the oceans.

Water Power

Islands are areas of land surrounded by water. As such, water often plays a large role in the existence of islands. The height of the surface of the ocean, or sea level, determines how much of a landmass appears above water. Rising ocean levels can create islands by flooding lower land between two taller landforms. The reverse process can cause two bodies of land to merge into a single one when lower land is uncovered.

The ocean is constantly eroding land. Waves crash on the shore, continually breaking rocks apart and washing sand away.

This never-ending process causes landmasses to shrink and sometimes disappear altogether. The same may happen to bodies of land in a river. As water rushes past land, it sweeps silt away downriver. This can reduce the size of an island. However, it can also create new islands at the end of a river as the silt being carried away is deposited in a new location. Wind and ice can have the same effect on land and also play an important role in the formation and destruction of islands.

CONTINENTAL AND SEDIMENTARY ISLANDS

One of Earth's largest landmasses, Australia, is technically the world's largest island. However, scientists categorize Australia as a continent and not an island because it is so large. Other large landmasses, such as Greenland, Iceland, and Madagascar, are considered islands. These landmasses share something in common with Australia. They were once joined together in the original landmass, Pangaea. Scientists call these kinds of islands continental islands.

Continental islands are islands lying near the coast of a continent. Many formed hundreds of millions of years ago and were separated from the mainland by geological forces. Others formed on or near continents much more recently as geological forces changed the shape and look of Earth's land and oceans.

What Is a Continental Shelf?

As mentioned in the previous chapter, Earth's crust is divided into seven major and many minor plates. Most continents rest on a single plate. The oceans also rest atop these plates. Earth's crust is much thinner at the bottom of the ocean than it is on the continents. The land

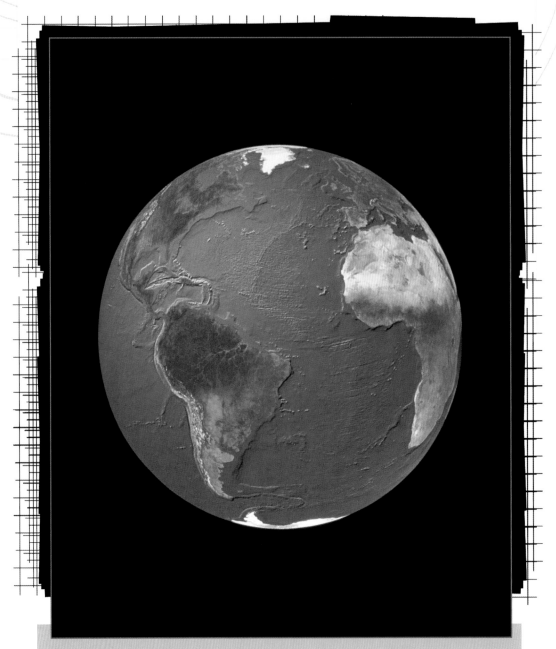

A topographic model of the Atlantic Ocean shows the relation between the continents and the tectonic plate boundaries. The Mid-Atlantic Ridge, faintly visible, runs through the center of the Atlantic Ocean.

between the shore of a continent and the ocean floor is called the continental margin. This area does not necessarily correspond to the location of plate boundaries.

The continental margin is divided into several regions. Just past the shore to any continent, Earth's crust slopes gently and gradually away, sometimes for hundreds of miles. This region is called the continental shelf, and it is where many continental islands can be found. The next region is the continental slope, which is a steeper drop to the ocean floor, or abyssal plain.

How Do Continental Islands Form?

All continental islands lie on the continental shelf in relatively shallow water—usually no deeper than 600 feet (183 meters). Very large, old continental islands, such as Madagascar off the east coast of Africa, broke away from the surrounding continents as they drifted apart. About 160 million years ago, Madagascar broke away from Africa, but it was still attached to India. Many millions of years later, India separated from Madagascar, leaving it about 250 miles (402 km) off the coast of Africa.

"Newer" continental islands, which may still be thousands of years old, were also once joined to the mainland but were separated in one of two ways.

Rising Sea Level

Many continental islands formed as the surface of the oceans rose to new heights. Earth's climate goes through cycles of warming and cooling. During the relatively cold part of the cycle, more of Earth's water is frozen at the poles. This causes the sea level to decrease, which means there is less water to cover low

Cadillac Mountain sits on top of Mount Desert Island in Acadia National Park, Maine. Visible as well is Bar Island.

areas near continental coasts. However, during the relatively warm part of the cycle, there is less ice at Earth's poles and more water in the oceans, which causes the sea level to rise.

When sea level rises high enough, the water flows onto low-lying land. This can create a new body of water between high coastal areas and areas farther inland. The high coastal areas then become continental islands. Decreasing sea levels can uncover lower land and reconnect continental islands with the mainland. This will certainly happen during the next ice age.

Tidal islands are extreme cases of islands that form as sea levels rise. What makes them different is that they usually form twice a day. Tides are the cyclic rising and falling of Earth's ocean surface, caused by the gravitational pull of the moon and, to a lesser extent, the sun. Tides rise and fall roughly twice a day. During low tide, shorelines are much wider than they are at high tide. During high tide, the

water reaches much farther up onto the shore. In some situations, small patches of land are connected to the mainland by a narrow land bridge. This land bridge can be natural or man-made. During high tide, these land bridges sink below sea level, creating a tidal island.

THE BRITISH ISLES

The British Isles are continental islands that used to be connected to the European mainland. The islands were formed by a combination of two forces: rising sea levels and sinking land.

During the last ice age, ice sheets covered much of the area that is today known as the United Kingdom. These ice sheets weighed heavily on the land, causing it to sink and creating lowlands and valleys. As Earth's temperature rose and the ice age came to an end, the sea level rose. The newly formed valleys and lowlands became flooded, creating parts of the North Atlantic and the English Channel. The higher lands remained uncovered with water and formed the British Isles.

Bar Island, part of Acadia National Park in Maine, is a tidal island. During low tide, visitors can walk—or drive an all-terrain vehicle—from Bar Harbor on Mount Desert Island to Bar Island across a natural land bridge. This natural bridge is called a bar. During high tide, the bridge becomes submerged, and visitors need a boat to reach Bar Island.

Sinking Land

Continental islands can form when dry land sinks into the ocean, leaving the higher land extending above sea level. Sometimes, a peninsula—a body of land surrounded by water on three sides, such as Florida—may become an island when the stretch of land closest to the mainland sinks, leaving the more distant land above the water. In other instances, tall hills or mountains remain above water as the lower land sinks. This can create a single island, or a group or chain of islands known as an archipelago. Archipelagos can form near continents, or they may be created by volcanoes far out at sea. Notable examples of continental archipelagos include Japan and the British Isles.

Sedimentary Islands

Sedimentary islands are similar to continental islands in that they form near the coasts of continents. They are the result of the erosion and movement of Earth's materials.

Barrier Islands

Thanks to ocean tides, the sand on a beach is continually being moved in one direction. A beach on one end of an island is often wider than the beach on the other end of the island. This is because waves pick up sand and move it farther down the beach.

Sometimes, waves deposit the sand they pick up farther out to sea. Over time, this sand builds up and forms underwater sand dunes or bars. Eventually, a sand bar can rise above the surface of the water, forming long, narrow islands called barrier islands that run parallel to the coast. Barrier islands can be 100 miles (161 km) or longer. They often occur in groups along a coast with spaces between them called inlets. Between the barrier island and the mainland is a small body of water. This body of water can be called a lagoon, bay, or sound depending on how big it is. Barrier islands get their name from the fact that they prohibit rough ocean waves from reaching the coast. They also provide protection for special ecosystems that develop on coasts and in the water between barrier islands and the coast. The Outer Banks off the coast of North Carolina and Fire Island off the coast of Long

Sand builds up on the ocean side of these barrier islands south of Long Island, New York. Fire Island is the long barrier island on the lower right..

Island, New York, are examples of barrier islands.

The Frisian Islands are a chain of barrier islands off the coasts of Germany, the Netherlands, and Denmark. Much like the British Isles, the Frisian Islands used to be connected to the mainland but were separated as ocean levels rose over the course of hundreds of years. About two thousand years ago, ocean water broke through coastal highlands and flooded the plain beyond. The islands began as rocks poking out of the ocean. Over the years, tides pushed more and more sand up against the rocks, forming ever-widening islands.

Delta Islands

As rivers rush down toward the sea, they pick up and carry bits of earth called silt. Silt is often carried many miles until a river reaches

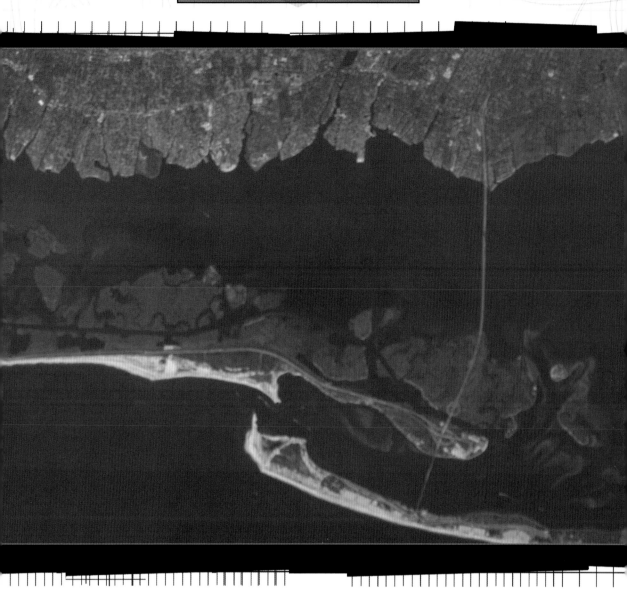

its end at a larger body of water. The river slows down as it reaches a lake or ocean and releases its grip on the silt, which then collects where the two bodies of water meet. Over time, the silt builds up and creates a landform called a delta.

Deltas are often fan-shaped. This shape is caused by the dispersal of silt as river water spreads out when emptying into a larger body of water. Deltas form as silt is deposited on the continental shelf near the end of a river. The silt can build up over the years and rise above the surface of the ocean. Islands may also form in or near a delta as water currents eat away at the land and change the way it looks. Delta land changes often due to shifting river speeds, ocean currents, waves, and weather. A delta island may not last long. The continued collection of silt may connect a delta island to the mainland, or ocean waves may carry it away.

Marajó Island, which lies at the end of the Amazon River in Brazil, is the world's largest fluvial, or river-created, island. It is 183 miles (295 km) long and 124 miles (200 km) wide, for a total area of 15,500 square miles (40,145 sq km). Marajó Island and the other islands of the Amazon delta were formed over hundreds of years as the Amazon River continually deposited silt where it meets the Atlantic Ocean. The delta channels around the island are constantly changing due to river currents and ocean tides.

Islands Formed by Glaciers

Glaciers are made of thick layers of ice and snow that accumulate over many thousands of years. Most glaciers can be found in mountain valleys. Gravity and pressure cause a glacier to "flow" very slowly over the land. As a glacier moves, it erodes the land beneath it and picks up sand and rock. This debris is often deposited on a coast as a glacier recedes, sometimes forming an island.

Long Island is the tenth-largest U.S. island. It was formed largely by a series of glaciers. As glaciers advanced south, they caused the land to bulge in front of them. As they receded, they deposited sand and rocks. Together, these two geological actions helped to build up the land that is today Long Island.

Glaciers also helped form the Great Lakes. First, they helped to "scoop out" the land to make lake beds and other landforms, including thousands of peaks that would eventually become islands. Second, melting glaciers helped fill the lakes. Manitoulin Island in Lake Huron is the world's largest freshwater island. The island itself has more than one hundred lakes, including Lake Manitou— the largest lake on the largest freshwater island in the world. Lake Manitou also has islands in it, making them islands in a lake on an island in a lake.

OCEANIC ISLANDS

Unlike continental and sedimentary islands, many islands form far out to sea where island-building forces are a bit different. Most oceanic islands are formed by underwater volcanoes far from the continental shelf.

Oceanic Rifts

Just as volcanoes form on continents where two tectonic plates meet, they also form on the bottom of the ocean. In all of Earth's oceans there are areas where two or more plates meet. Many of these plates are drifting away from each other. This forms a rift, or a long trench. As the plates continue to drift away from each other, magma rises to the surface. Lava is extruded from, or forced out of, the rift and spills away on either side, forming new oceanic crust. Eventually, the lava rises higher as new layers cover older layers. This forms a long, narrow stretch of land over the rift. This stretch of land is called a ridge.

In some cases, the ridge rises high enough to break the surface of the ocean, forming islands. The Mid-Atlantic Ridge runs down the center of the Atlantic

Ocean. Iceland, Bermuda, and the Azores archipelago were all formed by lava erupting from the Mid-Atlantic Ridge.

Island Arc

While some tectonic plates move away from each other, others move toward each other. When this happens, one plate is often forced beneath the other plate. This action is called subduction. The heat, pressure, and friction caused when this happens is great enough to create volcanoes on the upper plate. The volcanoes form a long line over the place where one plate is forced under another, an area called the subduction zone. This line of volcanoes is called a volcanic arc. On land, subduction zones result in the tallest mountains on Earth. The Rocky Mountains, the Himalayas, and the Andes all grew over subduction zones.

When subduction occurs at the bottom of oceans, underwater volcanoes rise up and eventually form an underwater mountain range. When the tops of these mountains break the surface of the water, a long line of islands called an island arc forms. The Aleutian Islands of Alaska are an example of an island arc. At the bottom of the North Pacific Ocean, the Pacific Plate is forced under the North American Plate, creating a subduction zone. A volcanic arc about 1,200 miles (1,931 km) long formed above the subduction zone. Approximately 300 of the volcanoes are tall enough to break the surface of the Pacific Ocean and create an island arc.

Mantle Plumes and Hot Spots

Many oceanic islands form over localized columns of hot mantle called mantle plumes. These columns may reach as deep as the

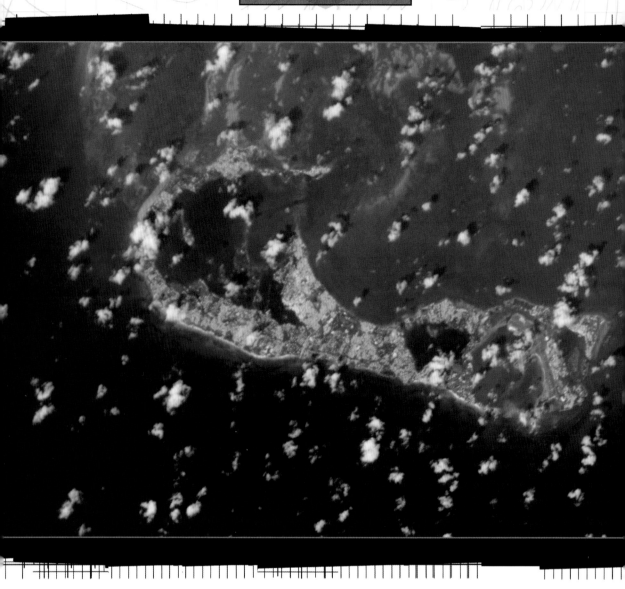

outer core and can last for hundreds of millions of years. Most mantle plumes have relatively short life spans—usually tens of millions of years. This may sound like a very long time to you, but when you take into account that Earth itself is about 4.5 billion

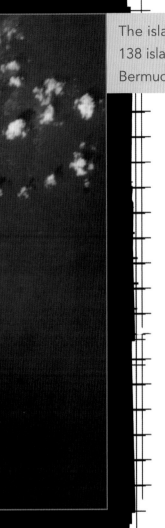

The island of Bermuda is one of a group of approximately 138 islands that form an archipelago. The total area of the Bermuda islands is 20.6 square miles (53.3 sq km).

years old, the life span of a mantle plume is like a fleeting moment.

The heat from a mantle plume rises continuously and causes the crust above it to swell and crack. Volcanoes often form where the crust cracks open. The area where this occurs is called a hot spot, for obvious reasons. Hot spots remain active for millions of years, constantly being fed fresh molten rock from the mantle plume below it. The lava that comes out of the hot spot spreads out as it cools. In time, the hardened layers of igneous rock may rise high enough to break the surface of the ocean. At this point, an oceanic island is born.

On the Move

Most archipelagos are chains of volcanic islands created over the course of millions of years by hot spots on the bottom of the ocean. As explained, volcanic islands form over hot spots in the ocean

Vast stretches of sea ice link the Aleutian Islands of Alaska. The archipelago of volcanic islands stretches for more than 1,200 miles (1,931 km).

floor, and the hot spots form over mantle plumes. Plumes with very long life spans are capable of creating long chains of islands. Mantle plumes stay relatively stationary for as long as they exist. A volcanic island may form above a hot spot. In time, however, the plate above the mantle plume drifts. The volcanic island will move away from the plume. A new hot spot forms directly behind the previous one over the mantle plume. This may result in the formation of a new volcanic island.

Over hundreds of millions of years, as the plate continues to float over the mantle plume, new hot spots continue to form along a path leading away from the original island. Eventually, this can form a long underwater mountain range. Many of the mountain peaks rise above the surface of the ocean, forming a chain of islands.

Older islands usually have little or no volcanic activity to help them to keep growing. After millions of years of erosion, they sink

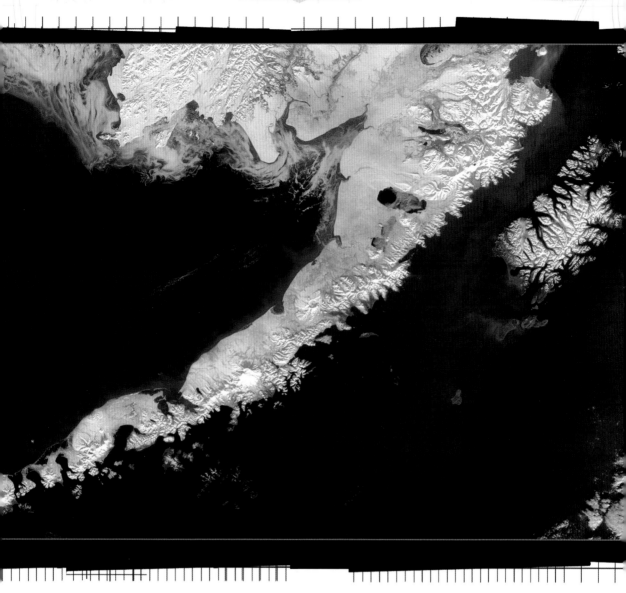

back into the ocean. The oldest often disappear as a plate they are on sinks below another plate. At the same time, on the other side of the island chain, a new island is being built up by fresh volcanic activity.

Mauna Loa, which is on the island of Hawaii, last erupted back in 1984. Many scientists have come to believe that the volcano may erupt again sometime during the next decade.

The Hawaiian Island Chain

Hawaii is one of the most well-known volcanic island chains on the planet. Hawaii is made up of eight main islands and many lesser islands made by hot spots that formed over a mantle plume in the Pacific Ocean. Generally speaking, the farther northwest one travels, the older the islands get. This is because the Pacific Plate is slowly moving northwest. As an island moves away from the mantle plume, another begins to form behind it. The island of Hawaii is the youngest and only volcanically active island. It is made up of five active volcanoes, including Mauna Loa, the tallest mountain in the world; measured from the bottom of the ocean to the volcano's peak, Mauna Loa is 33,000 feet (10,058 m) tall. Beneath the surface of the ocean to the southeast of Hawaii, an underwater volcano named Loihi is currently growing. One day, it may become an island in the Hawaiian island chain, too.

You might be surprised to hear that the Hawaiian Islands are just part of a much longer chain of mountains. The Hawaiian Islands and the Emperor Seamounts were all formed by the same mantle plume. The Hawaiian-Emperor Chain stretches about 3,600 miles (5,794 km) from Loihi to the Aleutian Trench. The chain has a bend in it because the direction of the Pacific Plate's movement changed millions of years ago. Older seamounts once existed, but they disappeared beneath the North American Plate long ago.

Coral Reefs and Atolls

Corals are small marine animals that live in shallow tropical waters. They protect themselves by forming a hard shell as they

The Great Barrier Reef is the largest coral reef in the world. Located just off the northeastern shore of Australia, it is actually made up of about 2,800 smaller coral reefs.

grow longer. Together, many corals help to produce beautiful underwater structures called coral reefs.

Coral reefs often build up around tropical volcanic islands. Corals must live in shallow water because they require sunlight to help make food. When a volcanic island ceases to be active, it often begins to erode and grow smaller. The coral reef around it will grow taller to remain near the surface of the water. Often, the coral grows above the surface of the water, forming a ring around the central volcanic island. In time, the volcanic island sinks below the surface of the water to become a seamount, leaving behind an island of coral surrounding a central body of water called a lagoon. This ring of coral is called an atoll. Since corals are living creatures, an atoll generally grows a little every year, but usually only by a few centimeters at a time. An atoll may be one circular island with a central lagoon or a ring of smaller islands around a lagoon.

It should be mentioned here that this is the most popular theory—first suggested by English naturalist Charles Darwin in 1842—to explain how atolls form, but there are others. Some scientists believe wind and ocean currents also help shape atolls. What is certain about the formation of atolls, however, is that it

WHERE IN THE WORLD IS SURTSEY?

The island of Surtsey is close to the southern coast of Iceland. It is just over 1 square mile (3 sq km), and its highest point is 560 feet (171 m) above sea level. Surtsey is a small, rocky island like many other volcanic islands all over the world, so what makes it so special?

Surtsey is one of the youngest volcanic islands on Earth. The formation of Surtsey began when an underwater volcano named Sutur broke open, allowing large amounts of lava to escape Earth's crust. Just like Iceland, Surtsey formed over the Mid-Atlantic Ridge. Over the next six months, the lava hardened to form new land. On November 15, 1963, Surtsey grew tall enough to break the surface of the ocean. After this occurred, great columns of ash rose into the air as lava continued to spill out of the volcano. At times, the column of ash could be seen in Reykjavík, the capital of Iceland.

The violent eruptions persisted for five months, followed by a long period of flowing lava that continually made the island bigger. This continued for the next three years until Surtsey fell silent in June 1967. Today, scientists continue to study Surtsey, which was designated a nature preserve in 1965. It continues to be the perfect "laboratory" for investigating lava, volcanoes, and the formation of new land.

takes thousands of years. Atolls are particularly low-elevation, fragile ecosystems. Should the sea level rise just a few feet, many of them would cease to be. Eventually, all atolls will sink below the surface of the ocean due to erosion and rising sea levels.

ISLAND NATIONS

A large majority of all the islands on Earth are inhabited. That means that they have people living on them. When you think of an inhabited island, you might think of a beautiful tropical location with lots of sun and sand. However, not all inhabited islands are tropical getaways. The following examples are just a few of the many island nations on Earth. Some certainly are tropical getaways, but some are less hospitable.

Greenland

About 65 million years ago, the land that is today known as Greenland was still fused within Pangaea. Many scientists believe that mantle plumes weakened the supercontinent at key areas, causing it to break into individual landmasses that drifted apart. Based on the presence of similar rocks, scientists know that Greenland was once attached to North America on its west coast and northern Europe on its east coast. Greenland is a continental island.

Greenland is the largest island on Earth, but it is also one of the most sparsely populated areas of our planet. It is 836,330.48 square miles (2,166,086 sq km), and has a

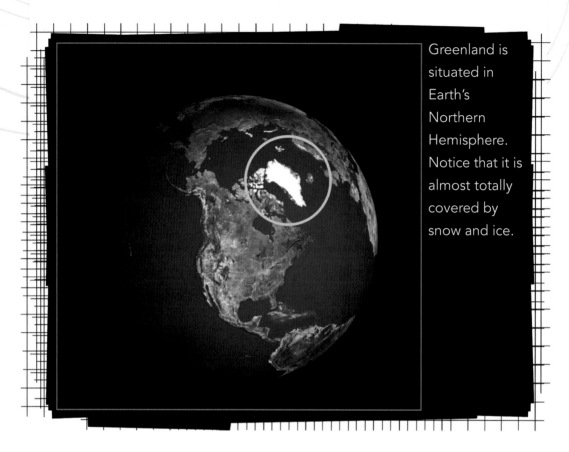

Greenland is situated in Earth's Northern Hemisphere. Notice that it is almost totally covered by snow and ice.

population density of about .03 people per square kilometer. The coast of Greenland—where most people live—is dominated by inlets and bays carved out by glaciers. The capital of Greenland, Nuuk, is the largest city. About 14,000 people live there.

Few people live in Greenland because much of it is covered by massive glaciers. The Greenland ice sheet covers about 80 percent of the island. The enormous weight of this ice sheet has caused the central land of Greenland to sink about 984 feet (300 m) below sea level. If the ice sheet were to melt, the sea level would rise enough to sink coastal cities like Los Angeles and New York City. Greenland would probably become an archipelago.

Iceland

About 200 miles (322 km) east of Greenland is the world's largest volcanic island—Iceland. The island has an area of about 39,768.5 square miles (103,000 sq km). The interior of the island is dominated by mountain peaks and ice fields, and few people live there. Most of Iceland's population (more than 304,000 people) lives along the coast, which is 3,088 miles (4,970 km) long thanks to numerous bays and fjords.

Iceland lies on the rift between two major plates—the Eurasian Plate and the North American Plate. The boundary between these plates, the Mid-Atlantic Ridge, runs along the bottom of the Atlantic Ocean. As the two plates on either side of the ridge move in opposite directions, magma rises and hardens to form new land along the ocean floor.

In addition to resting atop an ocean ridge, Iceland is also above a mantle plume. This is a somewhat rare occurrence that has resulted in a highly active volcanic island. Due to the high amount of volcanic activity, Iceland uses geothermal power to produce about one-quarter of the country's electricity. The volcanic activity also makes Iceland a "hot spot" for many geologists and volcanologists who want to study Earth's island-forming forces.

Japan

Japan is a large island arc formed over the area where the Pacific Plate and the Philippine Plate subduct, or are forced under, the Eurasian and North American Plates. Japan is situated on a long ring of volcanically active plate boundaries that scientists call the Ring of Fire, which encompasses the Pacific Ocean. Japan is an

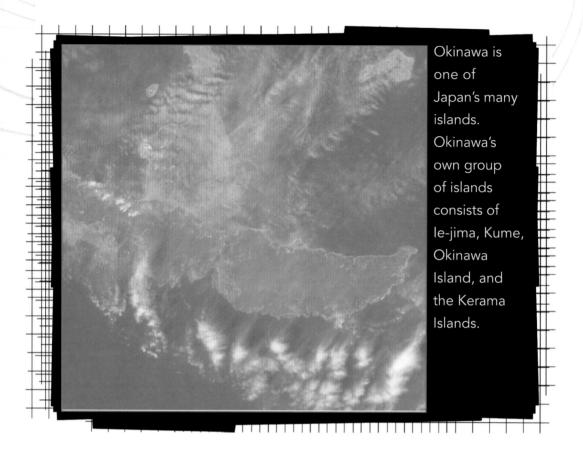

Okinawa is one of Japan's many islands. Okinawa's own group of islands consists of Ie-jima, Kume, Okinawa Island, and the Kerama Islands.

archipelago made up of more than 3,000 islands, including the four main islands of Honshū, Hokkaidū, Kyūshū, and Shikoku.

Millions of years ago, Japan was attached to Asia. As the Pacific and Philippine Plates were forced under the Eurasian and North American Plates, volcanoes formed over the subduction zone and the land began to rise. This action also caused a lower area to form behind the rising land, which often happens near a subduction zone. Japan began to drift eastward as the lower area widened. Eventually, the lower land flooded and became the Sea of Japan.

Because of its location on several fault lines, Japan is a volcanically active region. Mount Fuji, Japan's tallest mountain, is a volcano that last erupted about 300 years ago. Mountain chains dominate the four major islands, and earthquakes are common occurrences as Earth's plates continue to move. Hot springs in the mountains—heated by the same forces that create Japan's mountains and volcanoes—are a popular vacation spot for many Japanese vacationers. Because of the mountainous interior regions, the majority of Japan's population, about 127,500,000 people, lives in crowded cities along the coasts.

Samoan Island Chain

The Samoan Islands are a 350-mile (560 km) chain of islands in the South Pacific Ocean. They are divided into two different nations—Samoa to the west and American Samoa to the east. Nearly all of the people living on the island chain are of Polynesian descent.

Just as the Hawaiian Islands formed over a mantle plume, so did the Samoan Islands. At one time, many scientists believed that the Samoan Islands were an island arc formed as a product of subduction. South of the Samoan Island chain, the Pacific Plate is being forced under the Indo-Australian Plate. In addition to this, two of the westernmost Samoan islands, Savai'i and Upolu, have shown volcanic activity in the past century. This evidence led scientists to believe that the island chain was the result of subduction.

In 1995, earthquakes rumbling in the ocean east of Ta'u, the youngest of the Samoan Islands, gave scientists a clue that other geological forces were at work. Soon after, scientists discovered

NAURU

The world's smallest island nation, Nauru, is a volcanic island in the South Pacific Ocean. The island has an area of just over 8 square miles (21 sq km) and is one-tenth the size of Washington, D.C. The 18.5 miles (30 km) of coastline is surrounded by a coral reef. The interior of the island forms a plateau dominated by phosphate deposits. The tiny island is home to about 13,800 people. In 1999, Nauru became a member of the United Nations.

an active submerged volcano named Vailulu'u. They now believe this volcano rests atop the same mantle plume that created the other Samoan Islands. The scientists compared the igneous rocks of the new volcano with those of the westernmost Samoan Islands and discovered that they were made by the same mantle plume.

Vailulu'u is about 14,100 feet (4,300 m) tall. The top of the volcano is still 2,000 feet (600 m) below the surface of the ocean, but it is steadily growing. Some day in the distant future, it will become the newest Samoan Island.

The Maldives

The Maldives, also called the Republic of Maldives, is an island nation in the Indian Ocean southwest of India. The country is made up of an archipelago of 26 atolls. This includes about 1,190 islands, 200 inhabited islands, and 80 islands supporting tourist attractions. The country covers about 116 square miles (300 sq km)

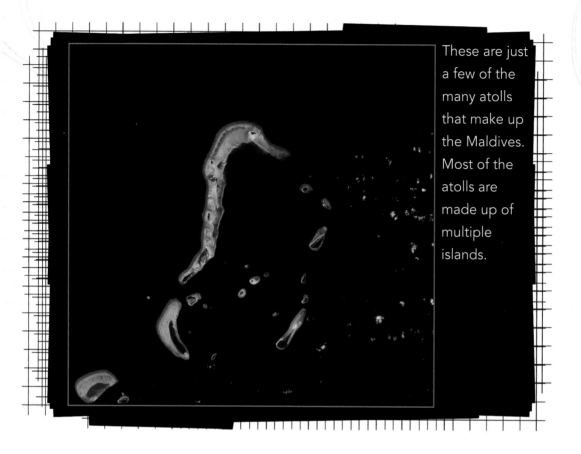

These are just a few of the many atolls that make up the Maldives. Most of the atolls are made up of multiple islands.

and has about 400 miles (644 km) of coastline. The highest point on the Maldives is only about 7.9 feet (2.4 m) above sea level.

Scientists studying the Maldives today believe the archipelago formed about 12,000 years ago as the sea level rose at the end of the last ice age. Although atolls are generally thought of as small, fragile locations, several of the Maldives atolls are large enough to support urban centers. Almost 380,000 people live there, but many more tourists visit the Maldives every year. Scientists believe that ocean currents and wind patterns help stir up nutrients that coral need for growth, which helps the Maldives remain healthy.

However, as world temperatures increase and polar ice melts, sea level is slowly rising. This could result in disaster for the Maldives. A rise in sea level of just 1 to 2 feet (.3 to .6 m) could cause great problems for hundreds of thousands of people living in the Maldives. Flooding would certainly drown the lower areas, but erosion would also cause extensive, irreversible damage to the islands. To battle the rising sea level, Maldivians have constructed concrete walls at strategic points along the coast, which helps to keep ocean waves from eroding the land. Scientists are currently working to come up with a plan to help save the Maldives from being taken over by the Indian Ocean.

MAN-MADE ISLANDS

Human beings are patient students and ingenious builders. For centuries, people have been making new islands— sometimes purposefully and sometimes by accident. The following section highlights several of the more interesting man-made islands.

Tenochtitlán

The Aztecs were a group of Native Americans who ruled a large portion of the land now called Mesoamerica from 1428 to 1521. The capital of the Aztec Empire was the city of Tenochtitlán, which was an island in Lake Texcoco. Today, this area is known as Mexico City. At its height, the city may have had between 200,000 and 300,000 people, which was larger than the populations of most European cities of the time.

The Aztecs arrived in Lake Texcoco around 1325. The early years in Tenochtitlán were very difficult because the island was marshy and not good for farming. Lake Texcoco was shallow and dirty. As the Aztecs grew in power over the next one hundred years, they added land to the island. They used wooden stakes tied together with reeds to make long,

This painting of Tenochtitlán was created by the well-known science-fiction artist John Berkey in 1983. Berkey presented a wide view of the city to give viewers an idea of its size and complexity.

narrow sections in the swampy water. They filled in these sections with dirt and plants to create floating garden beds they called chinampa, which comes from the Aztec word for "square made of canes."

Eventually, Tenochtitlán grew to 3.1 square miles (8 sq km). The Aztecs built a complex system of canals across the island. They also built three causeways leading to the mainland. One Spanish account reported that the causeways were wide enough for ten horses. The Spanish arrived in the early 1500s and destroyed the Aztec Empire. Unfortunately, they also drained Lake Texcoco, destroyed Tenochtitlán, and built Mexico City over the ruins.

The Floating Islands of the Uros Indians

Lake Titicaca is a lake nearly 12,500 feet (3,810 m) above sea level in the Andes of Peru. Prior to the arrival of Europeans, this area was home to many Native Americans. The Incas took control of much of the west coast of South America around 1438. Other tribes struggled to survive but were pushed farther away from the coast. The Uros Indians came up with a brilliant plan to avoid contact with the Incas and other enemies.

Abundant reeds grow along the banks of Lake Titicaca. The Uros learned to use the reeds for many purposes. They used them to make clothes, homes, and boats. They even used them to make floating islands. Pushed farther and farther away from the coast by stronger tribes, the Uros developed a method of weaving the reeds together to make thick, floating mats on which they could live. As the bottom of the mats rotted in the lake

water, new layers were added to the top to keep them afloat. The mats were strong enough to hold people, small animals, homes, and watchtowers. Some islands even had small holes cut into them for fishing.

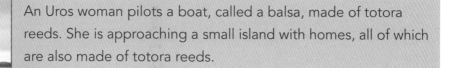

An Uros woman pilots a boat, called a balsa, made of totora reeds. She is approaching a small island with homes, all of which are also made of totora reeds.

Very few Uros are left today, but some of them still live on floating islands in Lake Titicaca. Today, you can travel to the Peruvian town of Puno and visit one of the last remaining Uros islands.

Belmont Island

Some artificial islands are the result of materials that were discarded in bodies of water during construction projects, such as subways. In the 1890s, piano manufacturer William Steinway wanted to build a trolley tunnel beneath the East River from Manhattan to his company town, Steinway, Queens. Construction began in 1892 on a small granite projection in the East River known as Man-o-War Reef. As workers dug into the granite, the debris they removed piled up around the worksite. Steinway died in 1896, but American financier August Belmont resumed work on the tunnel in 1907.

When the tunnel was finished, the debris removed from the shaft had created an artificial island measuring 100 by 200 feet

SEALAND: IS IT AN ISLAND?

During World War II, the United Kingdom constructed several military bases off the east coast of England for the purpose of defending the country against German aircraft and missiles. The bases were abandoned and destroyed after the war, except for one base—Fort Roughs Tower—which was built in international waters in the North Sea.

In 1967, former English military officer Paddy Roy Bates moved into Fort Roughs Tower with his family. With the help of his lawyers, Bates proclaimed the fortress a sovereign island and "micronation" called Sealand. He declared himself Prince of Sealand and his wife became a princess. Bates even had Sealandic coins minted.

The story of Sealand gets even more interesting. The micronation has been threatened by the English navy and terrorists. In 1978, Dutch terrorists came to Sealand and took Bates's son hostage. With the help of others, Bates retook the fortress and rescued his son, taking the terrorists prisoner in the process.

Today, Bates lives in Spain with his wife, and he leases Sealand to a "data haven" company named HavenCo. A data haven is a place or computer system that protects data from government action. People are still arguing about Sealand's status as an island and a micronation. Some claim it is a sovereign nation, while others say it is the property of England. What do you think?

(30.5 by 61 m). The official name of the island is Belmont Island, although many call it U Thant Island, named after a United Nations secretary-general from Myanmar. Today, the Steinway/ Belmont tunnel is part of the New York City subway system.

Subways pass beneath the artificial island several times a day. People are not allowed to visit Belmont Island because it is a protected sanctuary for migratory birds.

Islands in Dubai

The most ambitious man-made island project to date is underway in the city of Dubai, United Arab Emirates (UAE). The Palm Islands is the name for three island projects that began in 2001. Each island project entails the production of dozens of islands configured to resemble a palm tree.

Oil was discovered off the coast of Dubai in the 1960s. The UAE quickly became one of the world's wealthiest countries. However, the oil supply is dwindling quickly and will soon run out. Starting in the 1980s, leaders in Dubai sought to create a city that would attract travelers from around the world. The idea was to change the economy from one based on oil to one based on tourism. It quickly became the fastest-growing urban center in the world.

However, Dubai is located in the Arabian Desert with just 37 miles (60 km) of coastline on the Persian Gulf. By the early 1990s, hotels and high-rise buildings had formed a wall between the gulf and the hot, sandy interior. The ruler of Dubai, Mohammed bin Rashid Al Maktoum (Sheikh Mohammed), had a brilliant vision of how to increase Dubai's coast and ensure a robust, tourism economy.

In 1994, construction began on the Burj Al Arab, the tallest hotel in the world. This architectural masterpiece stands on an artificial island built in the Persian Gulf, 919 feet (280 km) away from the coast. It was designed to resemble a ship with tall sails.

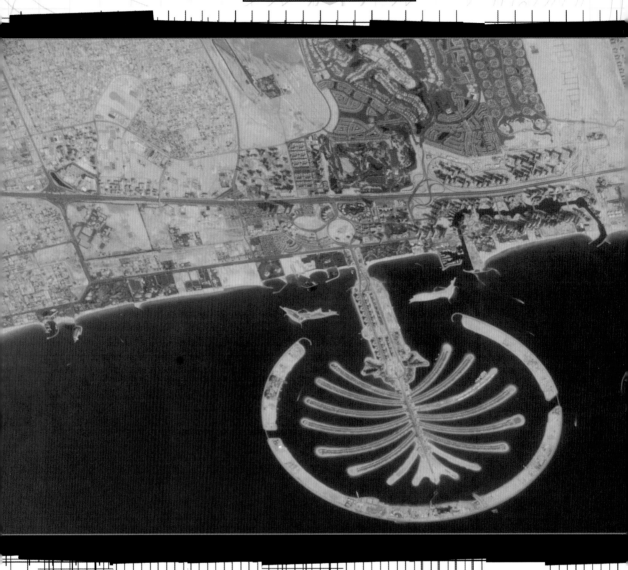

Just as Sheikh Mohammed had hoped, the iconic structure drew the attention of people from all over the world. Dubai's ruler hoped to take the success he had with the Burj Al Arab to an all-new level.

The Palm Jumeirah in Dubai, UAE, is the first and the smallest of the three Palm Islands. The islands feature more than thirty beachfront hotels and one floating hotel.

The Palm Islands

Today, Dubai is a modern marvel of architecture. It has the largest man-made harbor and an indoor ski slope. Soon, it will also have an underwater hotel and the world's tallest skyscraper. Perhaps the most amazing undertaking, however, is the construction of three artificial island projects collectively called the Palm Islands. The projects include the Palm Jumeirah, the Palm Jebel Ali, and the Palm Deira.

To make the Palm Islands, sand is dredged from the bottom of the gulf and is "sprayed" into place. This process is called land reclamation. The islands take shape as more sand is added. A special process is then used to pack the sand tightly to keep it from washing away. The centers of the three projects are designed to look like date palm trees. Barrier islands made of rock were built around the palm trees to keep them safe from wave erosion. As of the writing of this book, the construction of the Palm Islands is ongoing. As of 2007, many residents had moved onto Palm Jumeirah, although many of the resorts and shops are not yet finished.

The story of Dubai's waterfront does not end here. Future projects include the Dubai Waterfront, an ambitious project that will include artificial islands and canals in the form of an arc around Palm Jebel Ali. Construction has also begun on the World archipelago, which will consist of about 300 more man-made islands. Altogether, the man-made islands are expected to add about 600 miles (965.6 km) of coastline to Dubai.

The Human Influence

Earth's many islands—natural and man-made—are often at risk due to the actions of people. When people build on islands, they can cause erosion where it normally does not occur. This is especially harmful to beaches and barrier islands. Heavy tourism can have the same effect. Pollution caused by human industries can also have a devastating effect on Earth's islands. Island ecosystems can be permanently changed by air and water pollution. Many scientists believe that human pollution has already affected many islands by causing global warming. The hotter our planet gets, the more the polar ice caps will melt, and the higher our oceans will rise. If this continues to happen, many of Earth's islands will become submerged. Many other islands will be created by the same process as the sea level reaches new heights. Even though people are capable of making artificial islands—even some that can support thousands of people—we need to do what we can to protect the islands that are at risk.

GLOSSARY

debris The remains of something broken down or destroyed.

disaster An event that causes suffering or loss.

dredge To remove or recover material from under the water by dragging a net or frame along the bottom of a sea or river.

fault line The place on Earth's surface where two plates meet.

fjord A narrow bay surrounded by high cliffs.

geothermal power Energy created by heat stored in Earth's crust.

gravitational pull The force of massive bodies pulling objects toward their centers.

inhabited Having to do with an area where people live.

land reclamation The process of making land more useful or capable of sustaining life.

Mesoamerica A region of North America extending from Mexico down through Central America.

millennium A time span of one thousand years.

nutrient Any substance that provides a living thing with food.

phosphate One of a number of compounds in Earth's crust that contains the element phosphorous.

plateau A broad, high, flat piece of land.

population density A measurement of the number of people living in a unit of area.

sanctuary An area where wildlife is protected from human actions.

seamount A underwater mountain that does not break the surface of the ocean.

sediment Gravel, sand, or mud carried by wind or water.

sovereign To be free from outside control.

submerged Completely underwater.

trench A deep crack in the ocean floor.

tsunami A series of waves caused by a movement in Earth's crust on the ocean floor, volcanic eruptions, or landslides.

volcanologist A scientist who studies volcanoes.

FOR MORE INFORMATION

Geological Association of Canada (GAC)
Department of Earth Sciences
Room ER4063, Alexander Murray Building
Memorial University of Newfoundland
St. John's, NL A1B 3X5
Canada
(709) 737-7660
Web site: http://www.gac.ca
The GAC promotes the continued understanding and use of geosciences in public, professional, and academic life.

Geological Society of America (GSA)
P.O. Box 9140
Boulder, CO 80301-9140
(888) 443-4472
Web site: http://www.geosociety.org
The GSA provides resources crucial to earth scientists and fosters the continuous search for new information about Earth and its many ecosystems.

International Association of Volcanology and Chemistry of the Earth's Interior (IAVCEI)
Web site: http://www.iavcei.org
The IAVCEI promotes the study of volcanoes and volcanic processes, as well as coordinating the cooperation between scientists who study them.

U.S. Geological Survey (USGS)
12201 Sunrise Valley Drive
Reston, VA 20192
(703) 648-4000
Web site: http://www.usgs.gov
Per the USGS Web site, its mission is to "serve the Nation by providing reliable scientific information to describe and understand the Earth; minimize loss of life and property from natural disasters; manage water, biological, energy, and mineral resources; and enhance and protect our quality of life."

Web Sites

Due to the changing nature of Internet links, Rosen Publishing has developed an online list of Web sites related to the subject of this book. This site is updated regularly. Please use this link to access the list:

http://www.rosenlinks.com/lan/isla

FOR FURTHER READING

Arnold, Caroline. *Easter Island: Giant Stone Statues Tell of a Rich and Tragic Past*. New York, NY: Houghton Mifflin, 2004.

Blobaum, Cindy. *Geology Rocks: 50 Hands-on Activities to Explore the Earth*. Charlotte, VT: Williamson Publishing Company, 2008.

Hooper, Meredith. *The Island That Moved: The Forces That Shape Our Earth*. London, England: Frances Lincoln Children's Books, 2008.

Johnson, Rebecca L. *Plate Tectonics*. Minneapolis, MN: Twenty-First Century Books, 2006.

Kras, Sara Louise. *The Galapagos Islands*. New York, NY: Benchmark Books, 2008.

Stewart, Melissa. *Earthquakes and Volcanoes FYI*. New York, NY: HarperCollins Publishers, 2008.

Stewart, Melissa. *Extreme Coral Reef. Q & A*. New York, NY: HarperCollins Publishers, 2008.

BIBLIOGRAPHY

Central Intelligence Agency. "Maldives." *The World Factbook*, August 21, 2008. Retrieved August 27, 2008 (https://www.cia.gov/library/publications/the-world-factbook/geos/mv.html).

Central Intelligence Agency. "Nauru." *The World Factbook*, August 21, 2008. Retrieved August 27, 2008 (https://www.cia.gov/library/publications/the-world-factbook/geos/nr.html).

Dowdey, Sarah. "Why Is the World's Largest Artificial Island in the Shape of a Palm Tree?" HowStuffWorks.com, November 8, 2007. Retrieved August 26, 2008 (http://adventure.howstuffworks.com/dubai-palm.htm).

Fagin, Dan. "The Birth of Long Island." *Newsday*. Retrieved August 19, 2008 (http://www.newsday.com/community/guide/lihistory/ny-history-hs101a,0,4995000.story).

Florian, Jeff. "Dubai's Palm and World Islands—Progress Update." *AME Info*, October 4, 2007. Retrieved August 26, 2008 (http://www.ameinfo.com/133896.html).

Jennings, Terry. *Coasts and Islands*. North Mankato, MN: Smart Apple Media, 2002.

Knapp, Brian. *Landforms*. Danbury, CT: Grolier Educational, 2000.

Lippsett, Laurence. "Voyage to Vailulu'u: Searching for Underwater Volcanoes." Fathom.com. Retrieved August 27, 2008 (http://www.fathom.com/feature/122477/index.html).

Lutgens, Frederick K., and Edward J. Tarbuck. *Essentials of Geology*. Upper Saddle River, NJ: Prentice Hall, 2000.

Marshak, Stephen. *Earth: Portrait of a Planet*. New York, NY: W. W. Norton and Company, 2008.

Nationmaster.com. "Manitoulin Island." Retrieved August 27, 2008 (http://www.nationmaster.com/encyclopedia/Manitoulin-Island).

Nationmaster.com. "U Thant Island." Retrieved August 26, 2008 (http://www.nationmaster.com/encyclopedia/U-Thant-Island).

The Principality of Sealand. "History of Sealand." Retrieved August 26, 2008 (http://www.sealandgov.org/history.html).

PRNewswire. "Worlds of Discovery Planned for Nakheel's the Palm Jebel Ali in Dubai." February 28, 2008. Retrieved August 26, 2008 (http://www.prnewswire.com/mnr/seaworld/32017/).

Rothery, David A. *Geology*. Chicago, IL: Contemporary Books, 2003.

San Diego State University. "Krakatau, Indonesia (1883)." Department of Geological Sciences. Retrieved August 26, 2006 (http://www.geology.sdsu.edu/how_volcanoes_work/Krakatau.html).

Smith, Michael E. *The Aztecs*. Malden, MA: Blackwell Publishing, 2003.

SouthAmerica.cl. "Floating Islands—Peru." Retrieved August 26, 2008 (http://www.southamerica.cl/Peru/Floating_Islands.htm).

U.S. Geological Survey. "1883 Eruption of Krakatau." December 16, 2004. Retrieved August 26, 2008 (http://vulcan.wr.usgs.gov/Volcanoes/Indonesia/description_krakatau_1883_eruption.html).

Van Rose, Susanna. *Earth*. New York, NY: Dorling Kindersley, 2000.

Weier, John. "Amazing Atolls of the Maldives." Earth Observatory, NASA, May 1, 2001. Retrieved August 27, 2008 (http://earthobservatory.nasa.gov/Study/Maldives/maldives.html).

INDEX

About the Author

Greg Roza has written and edited educational materials for children for the past nine years. He has a master's degree in English from the State University of New York at Fredonia. Roza has long had an interest in scientific topics and spends much of his spare time tinkering with machines around the house. He lives in Hamburg, New York, with his wife, Abigail, and his three children, Autumn, Lincoln, and Daisy.

Photo Credits

Designer: Les Kanturek; Editor: Nick Croce
Photo Researcher: Marty Levick